THE METHYLENE BLUE BIBLE

An All-Encompassing Guide To Its Applications, History, Science, Contemporary Applications, Benefits, Advancements And Beyond

DR. KYREN STEVEN

Copyright © 2024 By Dr. Kyren Steven

All Rights Reserved...

Table of Contents

Introductory 4
CHAPTER ONE 6
   Uses And Applications 6
   Chemical Properties And Structure 10
   Antimicrobial Properties 14
CHAPTER TWO 19
   Antimalarial Uses 19
   Treatment Of Methemoglobinemia 23
CHAPTER THREE 29
   Biological And Cellular Effects 29
   Diagnostic Uses 34
   Side Effects And Adverse Reactions 39
CHAPTER FOUR 45
   Safe Handling Practices 45
   Environmental Applications 50
   Industrial Uses 55
   Conclusion 64
THE END 68

Introductory

Methylene Blue is a medication and dye that has several uses in medicine, biology, and chemistry:

- **Medicine**: It is used as a medication to treat methemoglobinemia, a condition where there is an abnormal amount of methemoglobin in the blood, which reduces the oxygen-carrying capacity of hemoglobin.

- **Microbiology**: Methylene Blue is used as a stain to visualize structures under a microscope. It can stain nuclei, bacteria, and other cell components, aiding in their identification and study.

- **Chemistry**: In chemistry, Methylene Blue is used as an indicator in redox titrations. It can change color depending on the oxidation state of the environment it is in, making it useful in determining the endpoint of titrations.

- **Aquariums**: It is also used in aquariums to treat fungal and bacterial infections in fish.

Methylene Blue works by converting methemoglobin back to normal hemoglobin, which can carry oxygen more effectively. It is typically administered intravenously in medical settings.

CHAPTER ONE
Uses And Applications

Methylene Blue has a variety of uses and applications across different fields:

Medical Uses:

- **Treatment of Methemoglobinemia**: Methylene Blue is used to treat methemoglobinemia, a condition where there is an abnormal amount of methemoglobin in the blood, which reduces its ability to transport oxygen. Methylene Blue helps convert methemoglobin back to normal hemoglobin, restoring oxygen-carrying capacity.

- **Diagnostic Aid**: In surgery and urology, Methylene Blue can be used as a dye to visualize anatomy and detect leaks in the urinary tract during cystoscopy.

Microbiological and Biological Uses:

- **Staining Agent**: Methylene Blue is commonly used as a biological stain in microscopy to visualize cells, nuclei, and other structures. It stains acidic structures blue and is used in both histology and microbiology for staining and differentiation of tissues and microorganisms.

- **Antimicrobial**: It has mild antifungal and antibacterial properties and is used in laboratories and sometimes in aquariums to treat fungal and bacterial infections.

Chemical Applications:

- **Redox Indicator**: Methylene Blue is used as a redox indicator in chemistry. It can undergo reversible oxidation-reduction reactions, changing color depending on whether it is oxidized (blue) or reduced (colorless). This property makes it useful in titrations to determine endpoints based on color change.

Photodynamic Therapy:

- In recent years, Methylene Blue has been explored for its potential use in photodynamic therapy (PDT), where it is activated by light to produce reactive oxygen species that can destroy cancer cells and bacteria.

Other Applications:

- **Veterinary Medicine**: Methylene Blue is used in veterinary medicine for similar purposes as in human medicine, such as treating methemoglobinemia in animals.

- **Aquarium Use**: It is used in fish tanks and aquariums to treat fungal and some bacterial infections.

Overall, Methylene Blue's versatility stems from its ability to interact with biological systems, its staining properties, and its redox characteristics in chemical applications. Its various uses highlight its importance in medicine, biology, and chemistry.

Chemical Properties And Structure

Methylene Blue is a heterocyclic aromatic chemical compound with the molecular formula $C_{16}H_{18}N_3SCl$. Here are its key chemical properties and structure:

Structure:

- Methylene Blue has a planar structure with three benzene rings fused together: two benzene rings are connected by a nitrogen and sulfur-containing bridge.

- The molecule has a central nitrogen atom that can exist in both oxidized (positively charged) and reduced (neutral) forms, giving it its redox properties.

Chemical Properties:

- **Redox Properties**: Methylene Blue is notable for its ability to undergo reversible oxidation-reduction reactions. In its oxidized form, it is blue (hence the name), and in its reduced form, it is colorless.

- **pKa**: Methylene Blue has a pKa value around 3.4, indicating it is a weak acid.

Solubility: It is soluble in water and forms a blue solution.

Applications of Redox Properties:

- In analytical chemistry, Methylene Blue is used as a redox indicator in titrations, where it changes color depending on the oxidation state.

- In medicine, its ability to accept electrons and donate them makes it effective in treating methemoglobinemia, where it reduces methemoglobin to hemoglobin.

Functional Groups:

Methylene Blue contains several functional groups:

- Aromatic rings (benzene rings) which contribute to its stability and conjugation.

- A central nitrogen atom bridged to a sulfur atom, which plays a crucial role in its redox behavior.

- Methylene (CH_2) group, giving the name "Methylene Blue".

Biological Interactions:

- Methylene Blue interacts with biological systems due to its ability to cross cell membranes and affect cellular functions.

- It can bind to DNA and RNA, affecting their structure and function, which contributes to its use as a biological stain in microscopy.

In summary, Methylene Blue's chemical structure and properties make it versatile in various applications, from medicine to analytical chemistry and biological sciences. Its unique redox behavior and solubility in water contribute to its widespread use and effectiveness in different fields.

Antimicrobial Properties

Methylene Blue possesses some antimicrobial properties, although they are generally considered mild and not as potent as some other antimicrobial agents. Here's an overview of its antimicrobial properties and applications:

Antifungal Activity:

- Methylene Blue has been shown to exhibit antifungal properties against certain fungal species. It can inhibit the growth of fungi such as Candida albicans, which is known to cause infections in humans.

- In laboratory settings and sometimes in clinical practice, Methylene Blue may be used as a topical treatment for fungal infections, although its effectiveness can vary.

Antibacterial Activity:

- Methylene Blue also exhibits mild antibacterial activity against various bacterial strains. It has been tested against bacteria such as Staphylococcus aureus and Escherichia coli.

- Its mechanism of action likely involves interfering with bacterial membranes or metabolic processes, although the exact mode of action is not fully elucidated.

Applications:

- **Microbiological Use**: In microbiology laboratories, Methylene Blue is sometimes used as a staining agent for bacterial cells and as a component in growth media to inhibit the growth of contaminants.

- **Clinical Use**: In clinical settings, Methylene Blue may be applied topically to treat localized bacterial or fungal infections. It has been used in wound care and certain dermatological conditions.

Limitations:

- While Methylene Blue has demonstrated antimicrobial activity, its effectiveness can be limited compared to dedicated antimicrobial agents like antibiotics or antifungal drugs.

- It is generally considered more suitable as an adjunct treatment or in cases where other options are not available or appropriate.

Photodynamic Therapy (PDT):

- In addition to its direct antimicrobial effects, Methylene Blue is also used in photodynamic therapy (PDT) protocols. In PDT, Methylene Blue is activated by light of a specific wavelength, leading to the generation of reactive oxygen species that can kill microbial cells, including bacteria and fungi.

While Methylene Blue does possess antimicrobial properties, they are typically used in specific applications such as microbiology, topical treatments for infections, and as part of PDT protocols. Its versatility stems from its ability to interact with microbial cells and its relative safety for use in certain clinical and laboratory settings.

CHAPTER TWO
Antimalarial Uses

Methylene Blue has shown potential in antimalarial treatment, although its use in this context is not as common or widely accepted as other antimalarial drugs like chloroquine or artemisinin-based therapies. Here are some aspects related to Methylene Blue's antimalarial uses:

Historical Context:

• Methylene Blue was one of the earliest synthetic drugs investigated for its antimalarial properties. It was first studied for this purpose in the late 19th and early 20th centuries.

Mechanism of Action:

• Methylene Blue's antimalarial activity is believed to involve several mechanisms, including inhibition of the electron transport chain in the parasite's mitochondria and interference with hemozoin formation (a byproduct of hemoglobin digestion by the parasite).

Clinical Trials and Research:

• Research on Methylene Blue as an antimalarial agent has been sporadic and has shown variable results. Some studies have demonstrated its efficacy against malaria parasites, while others have not shown consistent or sufficient efficacy to support widespread clinical use.

• Recent research has focused on combination therapies involving Methylene Blue with other antimalarials or as part of photodynamic therapy approaches.

Usage and Administration:

• In clinical settings, Methylene Blue has been administered orally or intravenously for the treatment of malaria. Intravenous administration may be used in severe cases or when other treatments are not available.

• It has also been used in combination with other antimalarial drugs to enhance effectiveness and reduce the risk of resistance development.

Challenges and Limitations:

- Despite its potential benefits, Methylene Blue faces challenges such as potential toxicity at higher doses and the emergence of resistance in malaria parasites.

- Its use as a primary antimalarial therapy is limited by the availability of more effective and safer drugs, especially in regions where malaria is endemic.

Photodynamic Therapy (PDT):

- In addition to its direct antimalarial activity, Methylene Blue is used in photodynamic therapy (PDT) for malaria. In this context, Methylene Blue is activated by light to generate reactive oxygen species that can kill malaria parasites in the blood.

While Methylene Blue has demonstrated antimalarial properties and continues to be studied for its potential in malaria treatment, its use is generally limited to specific cases and research contexts. Further studies are needed to clarify its efficacy, safety profile, and role in combating malaria, particularly in regions where effective antimalarial drugs are critically needed.

Treatment Of Methemoglobinemia

Methemoglobinemia is a condition where hemoglobin in the blood is oxidized to methemoglobin, which cannot bind and transport oxygen effectively. Methylene Blue is an effective treatment for methemoglobinemia, particularly when the condition is caused by certain drugs or chemicals that oxidize hemoglobin.

<u>Here's how Methylene Blue is used in the treatment of methemoglobinemia:</u>

Mechanism of Action:

- Methylene Blue acts as a reducing agent. It is converted to leukomethylene blue (reduced form) in the body, which then transfers electrons to methemoglobin, converting it back to normal hemoglobin (which is capable of binding oxygen).

- The reduction process converts methemoglobin (Fe^{3+}) to hemoglobin (Fe^{2+}), restoring the oxygen-carrying capacity of the blood.

Administration:

- Methylene Blue is typically administered intravenously in a hospital setting. The dose depends on the severity of methemoglobinemia and the patient's condition.

- It is given slowly over several minutes to hours, depending on the rate of response and monitoring of methemoglobin levels in the blood.

Indications:

- Methylene Blue is indicated for treating acquired methemoglobinemia caused by certain drugs (such as benzocaine, lidocaine, nitrites) or chemicals that oxidize hemoglobin.

- It may also be used in cases of hereditary methemoglobinemia if other treatments are ineffective or unavailable.

Monitoring:

- During treatment, methemoglobin levels in the blood are monitored regularly to ensure that they are reduced to safe levels.

- Response to treatment is assessed based on clinical improvement and oxygen saturation levels in the blood.

Adverse Effects and Considerations:

- Methylene Blue is generally well-tolerated when used appropriately for the treatment of methemoglobinemia.

- Adverse effects can include transient urine discoloration (blue or green), nausea, vomiting, and rarely, allergic reactions or hemolysis (breakdown of red blood cells).

Special Considerations:

- In cases where methylene blue is contraindicated (e.g., G6PD deficiency), alternative treatments such as exchange transfusion may be necessary.

- Pediatric dosing and administration may differ from adults and require careful consideration by healthcare providers.

Methylene Blue is a crucial and effective treatment for methemoglobinemia by reversing the oxidation of hemoglobin and restoring its ability to carry oxygen. Its administration requires careful monitoring and consideration of potential adverse effects, but it remains a cornerstone therapy for this condition.

CHAPTER THREE
Biological And Cellular Effects

Methylene Blue exerts various biological and cellular effects due to its interactions with biological systems. Here are some key effects and mechanisms of action:

Redox Reactions:

• Methylene Blue acts as a redox molecule, capable of accepting and donating electrons. This property is central to its biological effects, including its role in treating methemoglobinemia and its potential in photodynamic therapy.

• In cells, Methylene Blue can participate in redox reactions that influence cellular metabolism and oxidative stress pathways.

Mitochondrial Function:

- Methylene Blue has been shown to interact with mitochondrial respiratory chain complexes, particularly complex IV (cytochrome c oxidase). By enhancing electron transfer at this complex, it can improve mitochondrial function and ATP production.

- This property has been explored in neurobiology and aging research, where Methylene Blue has shown potential in enhancing mitochondrial function and reducing oxidative stress.

Antioxidant Properties:

- As a redox-active compound, Methylene Blue exhibits antioxidant properties. It can scavenge reactive oxygen species (ROS) and protect cells from oxidative damage.

- This antioxidant activity contributes to its potential neuroprotective effects and its role in reducing oxidative stress in various cellular contexts.

Neurological Effects:

- Methylene Blue has neuroprotective properties and has been studied for its potential in treating neurodegenerative diseases such as Alzheimer's disease and Parkinson's disease.

- It can modulate neuronal function and protect against neuronal damage induced by oxidative stress and protein aggregation.

Cellular Staining:

- In biological and histological applications, Methylene Blue is used as a stain to visualize cellular structures under microscopy. It binds to acidic components of cells, such as nuclei and granules, allowing for clear visualization and identification of cellular morphology.

Antimicrobial Activity:

- Methylene Blue exhibits mild antimicrobial activity against bacteria and fungi. It can disrupt microbial membranes and interfere with metabolic processes in microorganisms.

Photodynamic Therapy (PDT):

- In PDT, Methylene Blue is used in combination with light to generate reactive oxygen species (ROS). These ROS can induce cell death in targeted tissues, making PDT a potential treatment for certain cancers and infections.

Methylene Blue's biological effects stem from its ability to participate in redox reactions, modulate cellular metabolism, and protect against oxidative stress. Its diverse applications in medicine, biology, and research highlight its versatility and importance in understanding and treating various biological processes and diseases.

Diagnostic Uses

Methylene Blue has several diagnostic uses across different fields, primarily due to its staining properties and interactions with biological materials. Here are some key diagnostic uses of Methylene Blue:

Microscopy and Histology:

• Methylene Blue is commonly used as a biological stain in microscopy and histology to visualize cellular structures and components. It stains acidic structures such as nuclei and granules, providing contrast that helps in identifying cellular morphology and organization.

- In microbiology, Methylene Blue can be used to stain bacteria and other microorganisms, aiding in their identification and classification.

Cell Viability Assays:

- In cell biology and cytology, Methylene Blue can be used in cell viability assays. It stains viable cells blue, allowing researchers to distinguish between live and dead cells under a microscope or through colorimetric assays.

- The dye exclusion method using Methylene Blue is a simple and quick assay to assess cell viability based on the ability of live cells to exclude the dye.

Nuclear Staining in Research:

• In research settings, Methylene Blue is utilized for nuclear staining in various experiments. It can label nuclei in tissue samples or cultured cells, facilitating studies on cell proliferation, morphology, and differentiation.

Diagnostic Aid in Surgery:

• During surgical procedures, Methylene Blue can be injected or applied to visualize anatomy and detect leaks in the urinary tract or gastrointestinal system. It can help surgeons identify abnormalities or confirm the integrity of surgical repairs.

Detection of Nerve Damage:

- Methylene Blue has been used in diagnostic procedures to detect nerve damage. It can be injected near nerves, and its diffusion pattern can indicate nerve integrity or injury, aiding in surgical planning and treatment decisions.

Assessment of Mucosal Integrity:

- In gastroenterology and urology, Methylene Blue can be used to assess mucosal integrity. It stains the mucosal layer of tissues, helping clinicians detect lesions, erosions, or abnormalities during endoscopic examinations.

Dye Exclusion Tests:

- Methylene Blue is used in various dye exclusion tests in microbiology and cell biology. For example, it is used in the Wright-Giemsa stain for blood smears to differentiate blood cell types and identify abnormalities.

Overall, Methylene Blue's staining properties and ability to interact with biological structures make it a valuable tool in diagnostics, microscopy, and research. Its application spans from basic laboratory techniques to clinical procedures, aiding in the visualization and assessment of various biological and pathological conditions.

Side Effects And Adverse Reactions

Methylene Blue, while generally considered safe when used appropriately for its intended purposes, can cause certain side effects and adverse reactions, particularly at higher doses or with prolonged use. Here are some potential side effects and adverse reactions associated with Methylene Blue:

Common Side Effects:

• **Discoloration**: Methylene Blue can cause discoloration of the urine, turning it blue or green. This discoloration is harmless and temporary, typically resolving after discontinuation of the medication.

• **Nausea and Vomiting**: Some individuals may experience gastrointestinal symptoms such as nausea and vomiting, especially if Methylene Blue is administered at higher doses.

Rare but Serious Adverse Reactions:

• **Methemoglobinemia**: Although Methylene Blue is used to treat methemoglobinemia, excessive or rapid administration can potentially worsen methemoglobinemia or induce methemoglobinemia in susceptible individuals.

• **Hemolysis**: In individuals with glucose-6-phosphate dehydrogenase (G6PD) deficiency, Methylene Blue can trigger hemolysis (breakdown of red blood cells), leading to anemia and other complications.

• **Serotonin Syndrome**: In rare cases, Methylene Blue can interact with certain medications (such as selective serotonin reuptake inhibitors, monoamine oxidase inhibitors) and precipitate serotonin syndrome, a potentially life-threatening condition characterized by agitation, confusion, rapid heart rate, high blood pressure, muscle rigidity, and seizures.

Allergic Reactions:

• Allergic reactions to Methylene Blue are rare but can occur. Symptoms may include rash, itching, swelling (especially of the face/tongue/throat), severe dizziness, and difficulty breathing. Immediate medical attention is necessary if any allergic reactions occur.

Local Reactions:

- When used for diagnostic purposes or topical applications, Methylene Blue can cause local irritation or hypersensitivity reactions at the site of application.

Interactions:

- Methylene Blue can interact with other medications and substances. It should not be administered concurrently with serotonin-related medications (as mentioned above) or drugs that can induce methemoglobinemia (e.g., nitrites).

- Caution is also advised when combining Methylene Blue with certain anesthetics, as it can potentiate their effects and lead to adverse reactions.

Special Populations:

• Pregnant women and breastfeeding mothers should use Methylene Blue with caution, as its safety during pregnancy and lactation has not been fully established.

While Methylene Blue is generally safe when used appropriately under medical supervision, healthcare providers must carefully weigh its benefits against potential risks, especially in patients with underlying health conditions or specific genetic predispositions (such as G6PD deficiency).

Monitoring for adverse effects and adjusting treatment accordingly are essential aspects of safe Methylene Blue use.

CHAPTER FOUR
Safe Handling Practices

Safe handling practices for Methylene Blue are essential to minimize risks associated with its use, especially in clinical, laboratory, and industrial settings. Here are key guidelines for safe handling:

Personal Protective Equipment (PPE):

• Wear appropriate PPE, including gloves, lab coat or gown, and safety goggles or face shield when handling Methylene Blue. This helps protect against direct contact with the skin, eyes, or mucous membranes.

Ventilation:

- Use Methylene Blue in a well-ventilated area to prevent inhalation of vapors or aerosols. If working with powdered forms, ensure adequate local exhaust ventilation to control airborne dust.

Storage:

- Store Methylene Blue in tightly sealed containers in a cool, dry place away from direct sunlight and incompatible substances. Follow manufacturer recommendations for storage conditions.

Handling Precautions:

- Avoid direct skin contact with Methylene Blue. In case of accidental spills or splashes, promptly wash affected areas with soap and water.

- Use tools and equipment (such as spatulas, pipettes) dedicated for Methylene Blue handling to prevent cross-contamination.

- Ensure proper labeling of containers to indicate contents and potential hazards.

Administration and Dispensing:

- When administering Methylene Blue intravenously or in clinical settings, follow established protocols and guidelines for dosage, dilution, and administration rate.

- Use sterile techniques and equipment for preparing and administering Methylene Blue solutions to minimize the risk of contamination and infection.

Disposal:

- Dispose of Methylene Blue and contaminated materials (such as gloves, wipes, and empty containers) according to local regulations and institutional guidelines for hazardous waste disposal.

- Avoid disposal via sinks or drains unless permitted by applicable regulations.

Emergency Preparedness:

- Be prepared to respond to emergencies, such as spills, leaks, or accidental exposures. Have spill kits and appropriate cleanup materials readily available.

- Educate personnel on emergency procedures, including first aid measures and contacting medical professionals if exposure or adverse reactions occur.

Training and Awareness:

- Provide training to personnel involved in handling Methylene Blue on safe practices, potential hazards, and emergency procedures.

- Maintain awareness of updates in safety guidelines and recommendations related to Methylene Blue handling and use.

By adhering to these safe handling practices, healthcare professionals, laboratory personnel, and industrial workers can minimize risks associated with Methylene Blue and ensure its safe and effective use in various applications.

Environmental Applications

Methylene Blue has several environmental applications, primarily stemming from its properties as a dye and its ability to interact with organic matter and contaminants. Here are some key environmental applications of Methylene Blue:

Water Treatment:

- Methylene Blue can be used in water treatment processes, particularly in wastewater treatment plants. It is effective in adsorbing onto organic matter and certain pollutants, aiding in their removal from water.

- It can also be used in laboratory settings to detect and quantify pollutants and contaminants in water samples.

Indicator in Chemical Analysis:

- In environmental monitoring and chemical analysis, Methylene Blue serves as an indicator dye. It can be used to detect and measure the presence of reducing agents or to monitor redox reactions.

- Its color change from blue (oxidized form) to colorless (reduced form) makes it useful in titrations and analytical chemistry methods.

Soil Studies:

- Methylene Blue has been utilized in soil studies and agricultural research. It can help assess soil structure, porosity, and water-holding capacity by staining organic matter and soil particles.

Photodynamic Therapy in Environmental Applications:

- In recent years, Methylene Blue has been explored for environmental applications through photodynamic therapy (PDT) approaches. When activated by light, it can generate reactive oxygen species (ROS) that can degrade organic pollutants or pathogens in water or soil.

Microbiological Studies:

- Methylene Blue is used in microbiological studies related to environmental microbiology. It can stain bacteria and other microorganisms, aiding in their identification and classification in environmental samples.

Environmental Monitoring:

- In field and laboratory settings, Methylene Blue can be employed as a tracer dye to study water flow patterns, dispersion of pollutants, and the transport of contaminants in aquatic systems.

Research and Development:

- Researchers may use Methylene Blue in environmental research and development to study its interactions with pollutants, organic matter, and microbial communities in various environmental matrices.

It's important to note that while Methylene Blue has these potential environmental applications, its use must be carefully managed to prevent unintended environmental impacts. Proper disposal practices and adherence to regulatory guidelines are crucial to minimize any adverse effects on ecosystems and aquatic life.

Industrial Uses

Methylene Blue has several industrial uses, primarily centered around its applications as a dye and its chemical properties. Here are some key industrial uses of Methylene Blue:

Dyeing and Textiles:

- Methylene Blue is widely used as a dye in the textile industry to color cotton, silk, and wool fabrics. It produces a deep blue color and is also used as a biological stain in histology and microbiology.

Photography:

- In traditional photography, Methylene Blue has been used as a component in certain types of photographic toners and developers. It helps to intensify black and white prints.

Laboratory and Chemical Applications:

- Methylene Blue serves as a redox indicator in analytical chemistry. It undergoes reversible oxidation-reduction reactions, changing color depending on its oxidation state. This property makes it useful in titrations to determine endpoints based on color changes.

Medicine:

• While primarily used in medicine for its therapeutic properties (e.g., treating methemoglobinemia), Methylene Blue also finds niche applications in medical device sterilization and as a biological stain in laboratory settings.

Environmental Applications:

• As mentioned earlier, Methylene Blue has environmental applications, including its use in water treatment processes, soil studies, and environmental monitoring. It can adsorb onto organic matter and certain pollutants, aiding in their removal from water systems.

Research and Development:

• In research and development, Methylene Blue is used in various studies, including biochemical research, cellular biology, and pharmacology. Its ability to interact with biological systems and its redox properties make it valuable in exploring various biological processes and potential therapeutic applications.

Veterinary Medicine:

- Methylene Blue is used in veterinary medicine, similar to its applications in human medicine, such as treating methemoglobinemia in animals and as a biological stain in veterinary diagnostics.

Other Applications:

- In addition to the above, Methylene Blue may have applications in niche industries such as cosmetics (hair dyes), plastics (as a dye), and analytical chemistry (as a standard reference material).

In general, Methylene Blue's adaptability is a result of its dyeing properties, redox chemistry, and interactions with biological systems, which render it useful in a variety of industrial applications in addition to its well-known medical applications.

Disclaimer:

The information provided in this guidebook on Methylene Blue is for educational and informational purposes only. It is not intended as a substitute for professional medical advice, diagnosis, or treatment. Always seek the advice of your physician or other qualified health provider with any questions you may have regarding a medical condition or treatment.

Methylene Blue should only be used under the supervision of a healthcare professional. The authors and publishers of this guidebook are not liable for any direct, indirect, incidental, or consequential damages that may result from the use or misuse of the information contained herein. The reader assumes full responsibility for consulting a qualified healthcare professional regarding health conditions or concerns before starting any new treatment or discontinuing an existing treatment.

This guidebook makes no claims about the efficacy, safety, or appropriateness of Methylene Blue for any individual. It is essential to consider individual health needs and conditions when making medical decisions. The content provided is based on current knowledge and research as of the date of publication, but medical knowledge is constantly evolving, and new information may emerge that could alter the relevance, accuracy, or completeness of the information provided.

By using this guidebook, you acknowledge and agree that you have read, understood, and accepted this disclaimer.

Conclusion

Methylene blue is a chemical compound that is highly adaptable and has a diverse array of applications in a variety of disciplines. It is distinguished by its capacity to function as a therapeutic agent, biological stain, and redox indicator.

The following are the primary elements that encapsulate its significance and applications:

- Medical Applications: Methylene Blue is essential in the treatment of methemoglobinemia by converting methemoglobin back to hemoglobin. Additionally, it has the potential to be employed in diagnostic procedures and in photodynamic therapy.

- Biological and Microbiological Applications: It is used as a stain in microscopy and histology to facilitate the visualization of cellular structures. Furthermore, it is employed in biological research and possesses moderate antimicrobial properties.

• Chemical Properties: Its redox properties render it a valuable redox indicator in titrations and other chemical analyses, making it useful in analytical chemistry.

• Industrial and Environmental Applications: Methylene Blue is employed in various industries, including textiles (as a dye), water remediation, and photography. It also has applications in environmental research and studies, with a particular emphasis on the study of soil, water quality, and environmental contaminants.

• Safety and Handling: It is imperative to comply with proper handling practices to prevent potential side effects, including nausea, discoloration of urine, and uncommon but severe reactions such as methemoglobinemia or allergic responses.

• Future Directions: Methylene Blue is still being investigated in various fields, including neuroprotection, environmental remediation, and advanced therapies, as part of ongoing research.

Methylene Blue's multifaceted nature and beneficial properties render it a valuable tool in a variety of scientific, medical, and industrial settings, substantially contributing to environmental management, treatment, diagnostics, and research. Its significance in the advancement of science and the resolution of a variety of challenges across various disciplines is underscored by its ongoing exploration and utilization.

THE END

www.ingramcontent.com/pod-product-compliance
Lightning Source LLC
Chambersburg PA
CBHW071844210526
45479CB00001B/276